Graph Paper Designs

EASY AND CREATIVE ART WORK

JOHN H. LETTAU

INTRODUCTION TO GRAPH PAPER DESIGNS

This booklet will demonstrate how to use simple graph paper sheets in the creation of exciting and fascinating designs. Several important points should be noted and understood before you begin to work on suggested patterns. One way to make design construction easy is to free-hand all lines of your designs. Never use a straightedge or ruler. Free handing will become very easy over time and with practice.

Most student created designs found on pages 28-58 have a super-imposed section of graph paper in the upper left-hand corner. This allows you to see the construction pattern of each design. As in the Starting Lesson, page 4, there is a pattern in how the design is drawn find the pattern and all you have to do is repeat it. In the case of the Starting Lesson the pattern is square-triangle-triangle-square repeated in all directions.

Area, symmetry, geometric shapes, coordination, direction following, tessellation and creativity are a few of the related areas to be explored and taught while working on designs. The terms Straight-Line Pattern, Curved Line Patterns, Multi-Lined Pattern and Negative-Positive Pattern are use through the following pages. Below is a sample of the various design patterns that will be covered in detail.

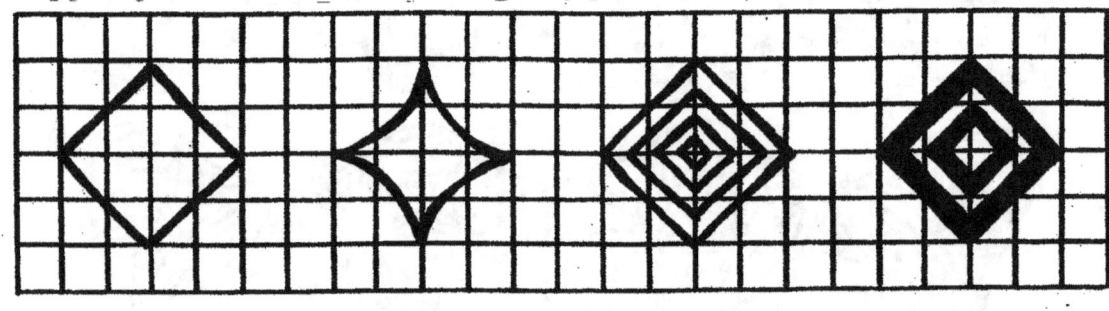

| Straight Line Pattern | Curved Line Pattern | Multi-lined Pattern | Negative-positive Pattern |

**Be Creative
Do Your Best Work
****
Anything Worth Having
is Worth Working For**

Imaginative Graph Paper Designs
Are Fun to Create
Take Pencil in Hand and Create
Some of Your Own
Graph Paper Designs
and

STARTING LESSON

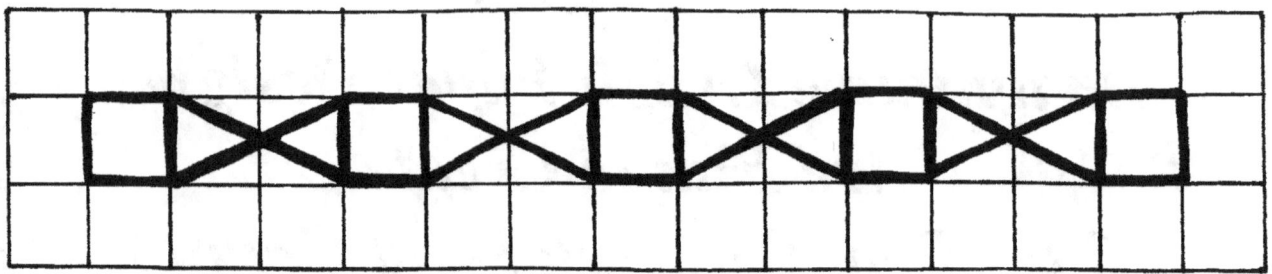

1. Place sheet of graph paper horizontally on desk top.
2. Freehand a square in the upper left-hand corner
3. Once again...DO NOT USE A RULER or STRAIGHTEDGE.
4. After the square draw 2 triangle as shown.
5. Continue square-triangle-triangle pattern to the right side of the paper.

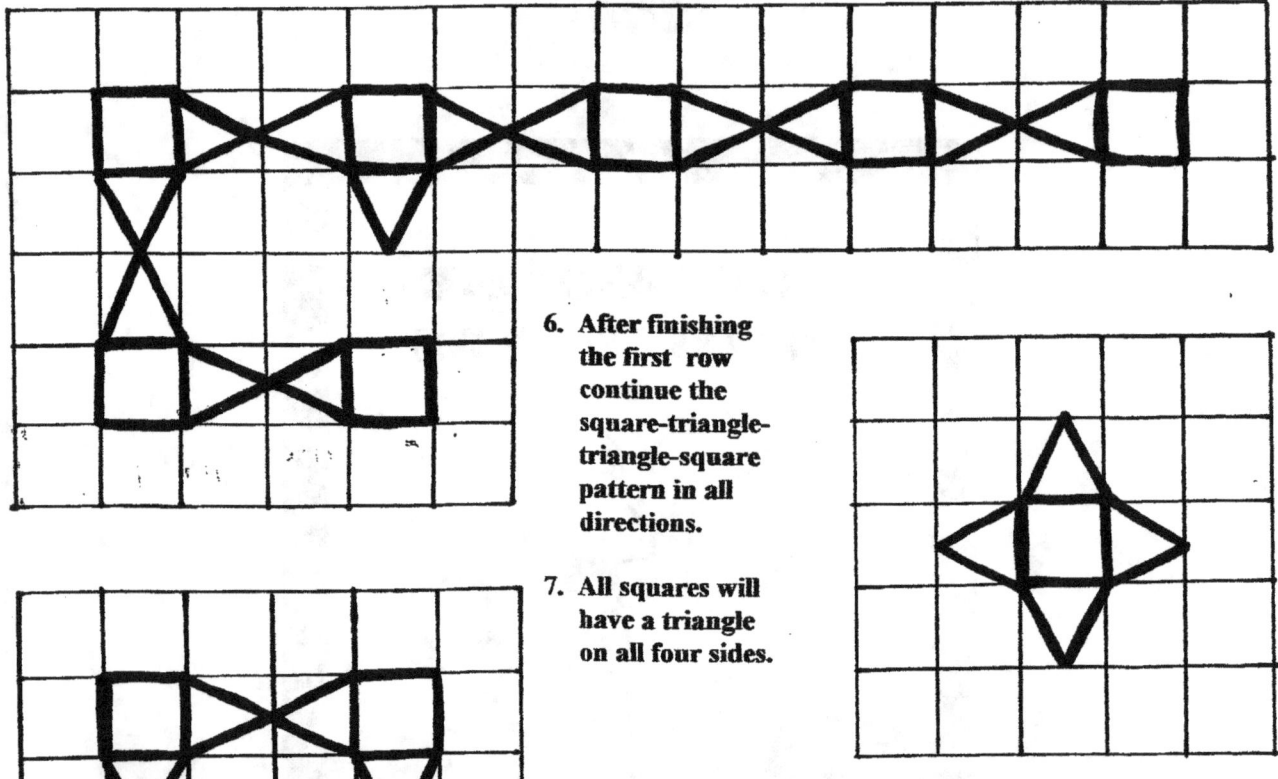

6. After finishing the first row continue the square-triangle-triangle-square pattern in all directions.

7. All squares will have a triangle on all four sides.

8. As you continue the basic square-triangle-triangle-square pattern in both the vertical and horizontal directions an octagon will be formed.

4

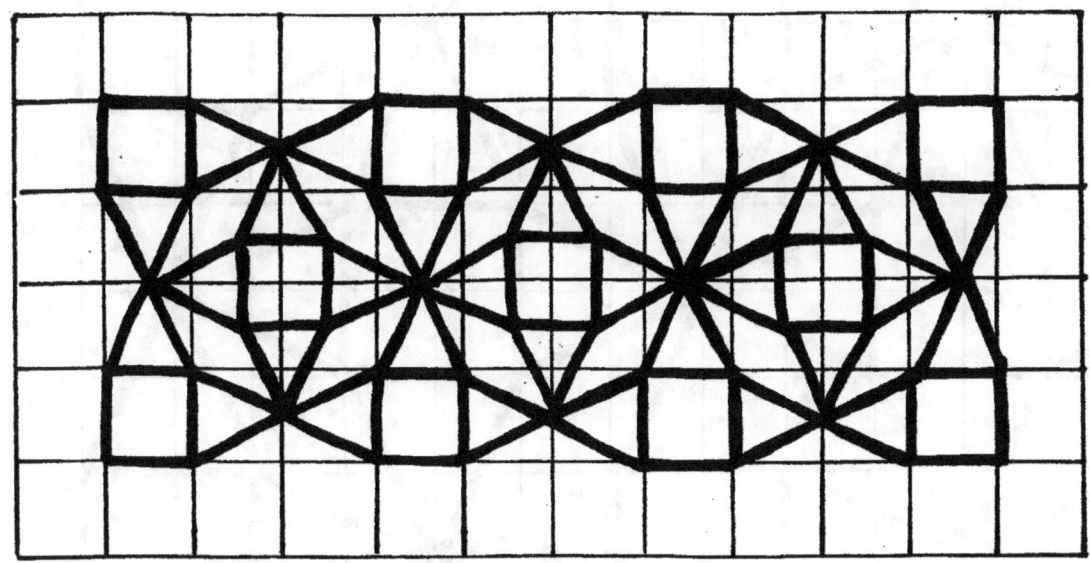

9. It is a simple matter to complete the design by filling in all octagons with squares and triangles as shown above.

10. Place a square in the center of the octagon. Draw a triangle on each to the four sides of the square. Notice that four diamonds have been formed.

11. Turn to page 6 for a completed design

12. To complete the design color all squares one color all triangles a second color and all diamonds a third color.

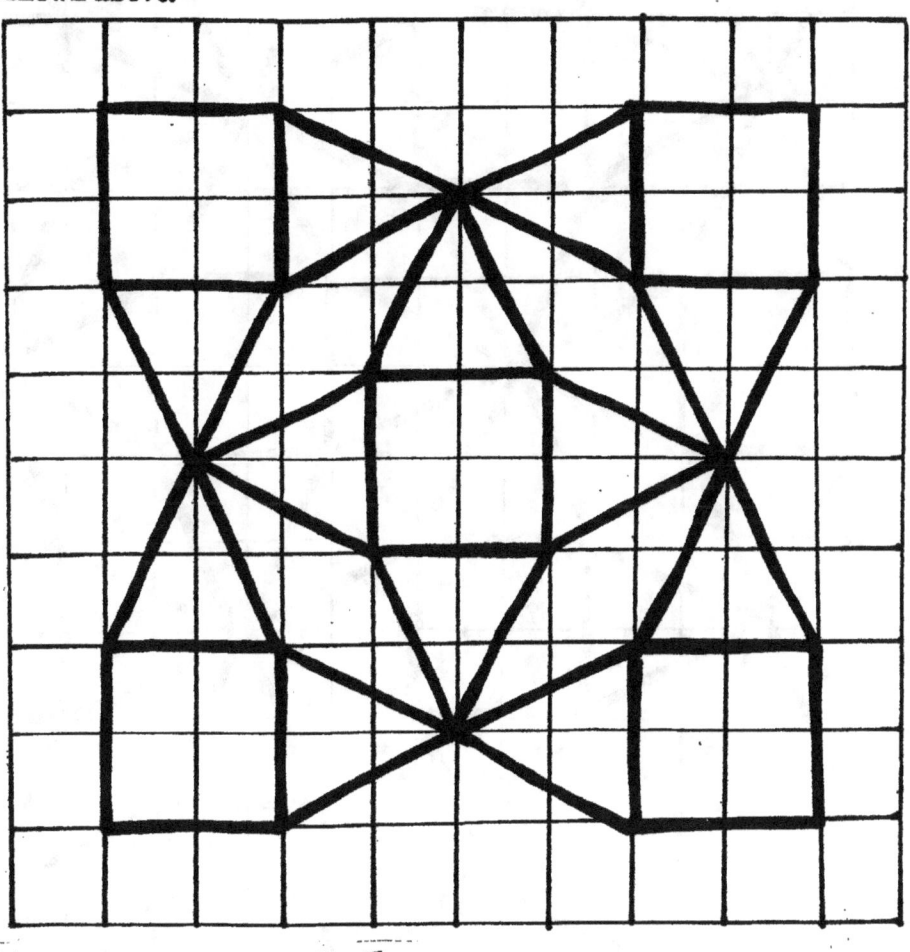

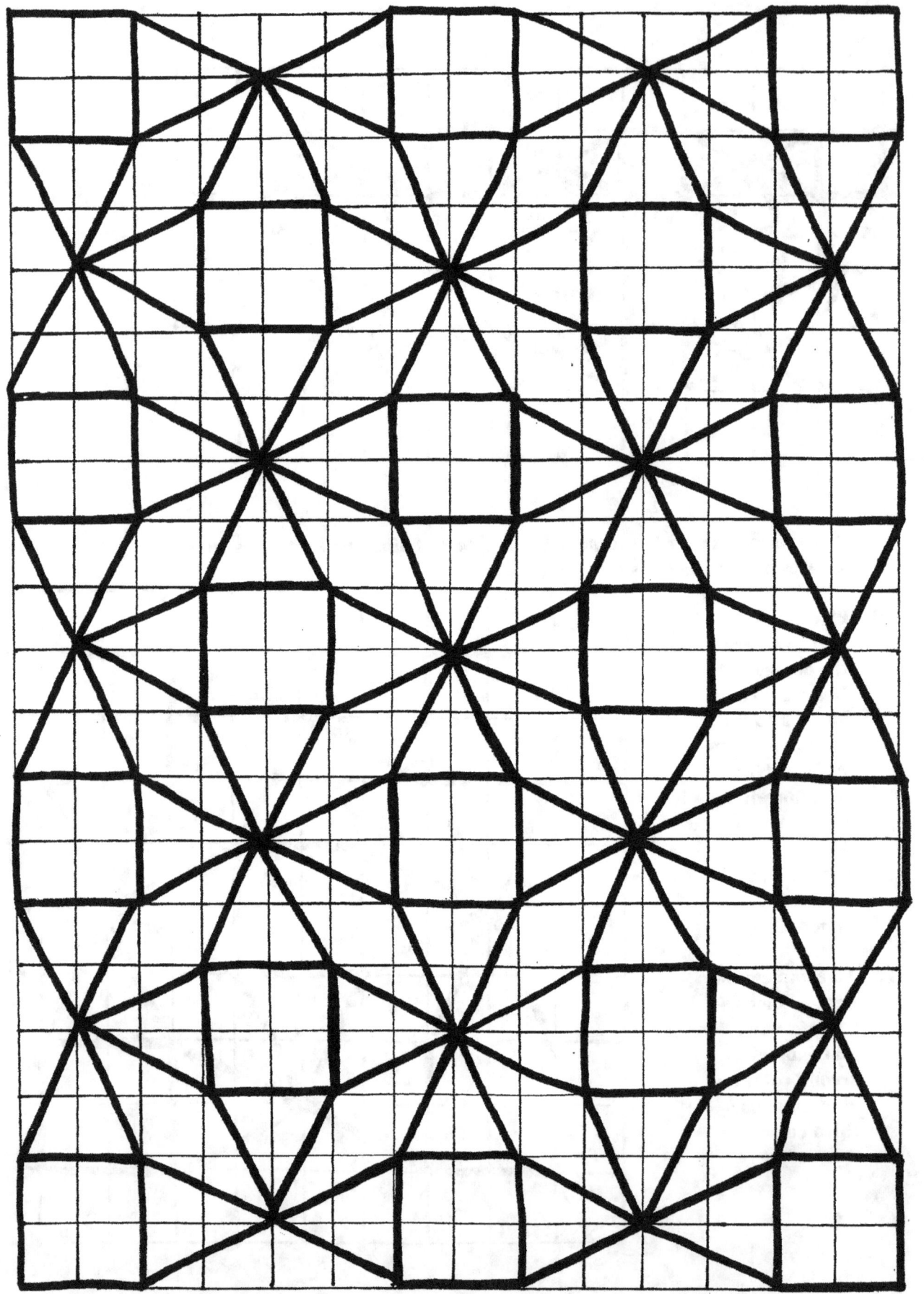

SQUARE LINE PATTERNS

Squares can be used in the construction of many different Graph Paper Designs. The squares shown are samples of different sizes that can be used when creating designs.

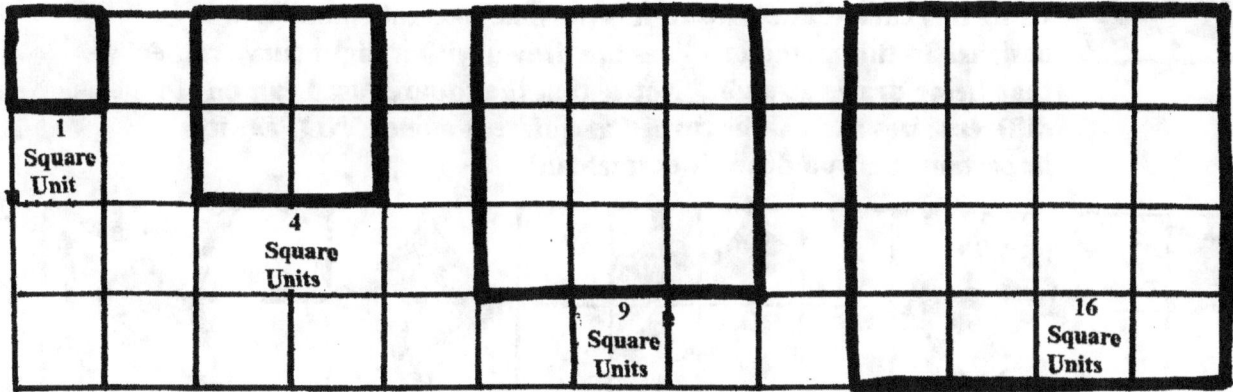

By drawing diagonal lines in the above squares we change the size and position of the square in the center of the original square. The size of the new square is half the area of the original square. The concept that the resulting square is ½ the area of the original is important when teaching area.

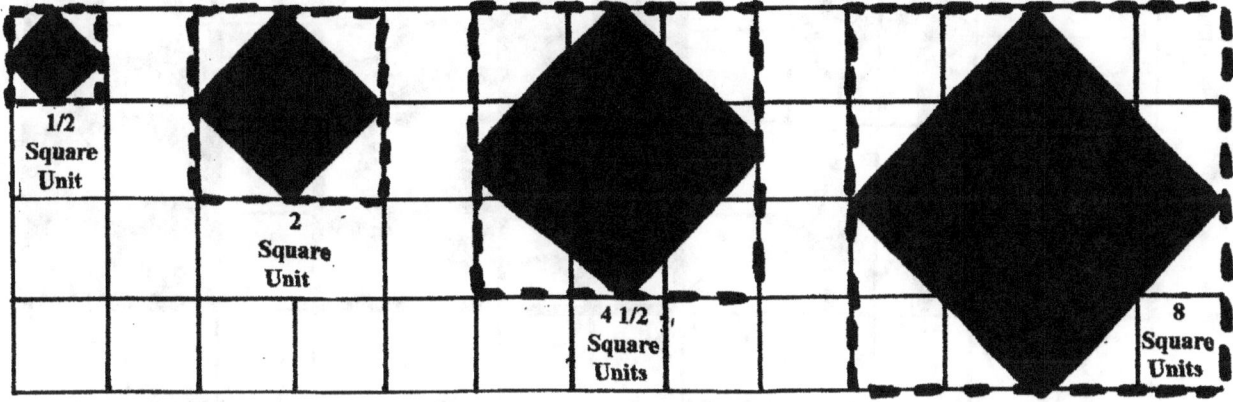

Examples using other square variations in the creation of new design patterns.

7

CURVED LINE PATTERNS

Using this curved line pattern it is possible to create many new designs. In this method all lines are drawn with a slight curve rather than being drawn straight, Notice that the squares used can be of different sizes and can be turned at different angles. NOTE...it is important that you draw lines freehand.

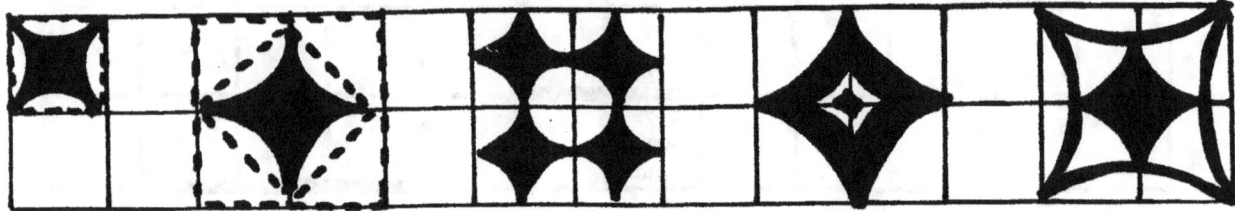

Examples using other variations in the creation of new design patterns. Try curved rectangle patterns in clusters and at different angles as the possibilities are endless and fun.

STRAIGHT & CURVED LINE PATTERNS

This design pattern is the result of the Starting Lesson on page 4. Note...there are only freehanded straight lines in this design. This is the basic Square-Triangle-Square that is the starting point of design creations in this book.

Here is the Starting Lesson Design using the curved lines method as shown on page 8. Note the difference in the two design styles. The curved line concept creates many different design options. The completed pattern is on page 10.

TRY BOTH DESIGN STYLES ON YOUR OWN CREATIVE CREATIONS

10

SQUARE PATTERN VARIATIONS

This square variation has lots of possibilities for creating geometric graph paper designs and teaching the concept of area. Note the large square has an area of nine square units and the smaller square has an area of five square units. Teach area as you progress through Graph Paper Designs.

A few changes to the above pattern and another pattern appears. A complete full page design can be found on page 12.

DIFFERENT SQUARE PATTERNS

12

PATTERNS USING BLOCK LETTERS

Use your imagination and creativity to form patterns from block letters.

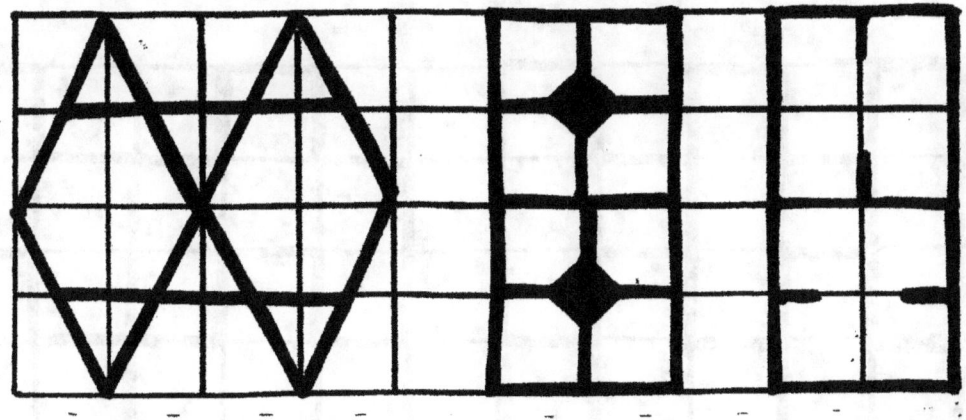

BLOCK LETTER A   BLOCK LETTER B   BLOCK LETTER C

BLOCK LETTER M   BLOCK LETTER X   BLOCK LETTER E

NOW TRY OTHER BLOCK LETTERS!!!!

TESSELLATION PATTERNS DESIGNS

Tessellation is the continued use of one shape in a given pattern. Use your imagination and creativity to create other tessellation patterns. Try using squares, triangles, hexagons, trapezoids, etc. Have fun creating designs.

15

16

MULTI-LINED PATTERN
&
NEGATIVE- POSITIVE PATTERN
DESIGNS

To produce the Multi-Lined Pattern draw your design and continue to design the same shape but continue to reduce the size as you go along. The number of lines you draw is up to you. Experiment to the select the style you like best. A full sheet using this pattern is located on page 19.

To produce the Negative-Positive Pattern do the same as for Multi-Lined Pattern. With the completed Multi-Lined Pattern you shade alternate spaces of the design with crayons or colored pencils. Try using one color and then try several different colors. A full sheet using this pattern is located on page 21.

MULTI-LINED PATTERN
&
NEGATIVE- POSITIVE PATTERN
DESIGNS

This is a small section of a Multi-Lined Pattern drawn on ½ inch square graph paper. The basic design is on page 4. A full page design is on page 19. To see the contrast when one square inch graph paper is used turn to page 20.

This is a small sectioon of the Negative-Positive Pattern drawn on ½ inch graph paper. Once again the pattern on page four is the basic design. Turn to page 21 for the completed design on ½ inch square grap paper. Page 22 is the same pattern using one inch square graph papper. Compare both patterns with the two designs on page 9. The basic pattern for all four is the Starting Lesson on page 4.

19

20

21

22

TEACHING AREA WITH DESIGNS

Notice how the squares in this diagram get larger and smaller. The addition of new square will go on to infinity in both directions. There is a definite number pattern.
A=128 sq. units   B= 64 sq. units   C=32 sq. units   D=16 sq. units   E= 8 sq. units
F=4 sq. units   G=2 sq. units   H=1 sq. unit   I=½ sq. unit   J=¼ sq. unit   K=?   L=?
What is the pattern???

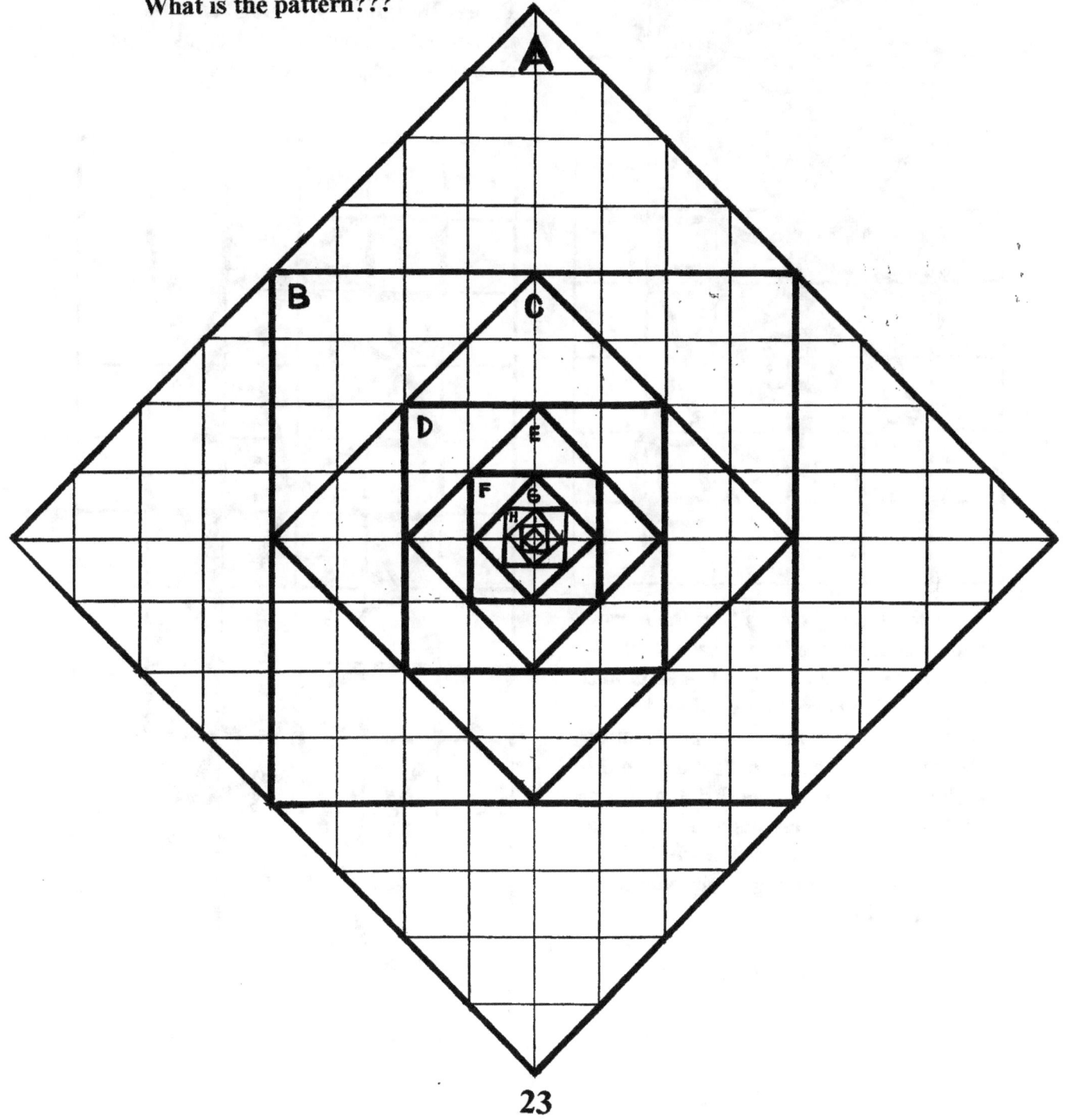

23

STUDENT & CLASS BOOKLETS

Consider projects that include the construction and design of various booklets of student created Graph Paper Designs. Booklets can be of individual student designs or classroom designs as a whole.

With the quality copies that the newer copy machine produce excellent black and white design copies. Booklets are very creative ways to record and store student and class Graph Paper Designs. A class collection could include as least one good design from each student. Students may want to make a collection of their own designs in addition to having their works of art in the class booklet.

Have a class contest to design the cover for a class booklet.

24

WAYS TO SHOW TEXTURE

To show texture on copies of student Graph Paper Designs that are ro be included in student and class booklets ...use the following suggestions. Texture can be shown using circles, diagonal lines, dots, shading...etc.

This sample textured design has all triangles shown with dots, all squares with shading and all diamonds with lines.

A full page design using this texture pattern is on the next page, page 26.

OTHER SUGGESTED TEXTURE PATTERNS

25

26

GRAPH PAPER DESIGN PATTERNS

SAMPLE GRAPH PAPER MASTERS

On pages 59-61 are several different size sample graph paper patterns to include ¼ inch, ½ inch and 1 inch squares.

DESIGN LESSON VARIATIONS

Use class completed Starting Lesson Design, page 4, as the background for a bulletin board.

Make a booklet of the student created designs.

After class has had time to create designs following direction for the basic design, page 4, have students do the starting design with the following...Multi-Lined pattern, Negative- Positive Pattern and the Curved Line Pattern.

28

29

30

31

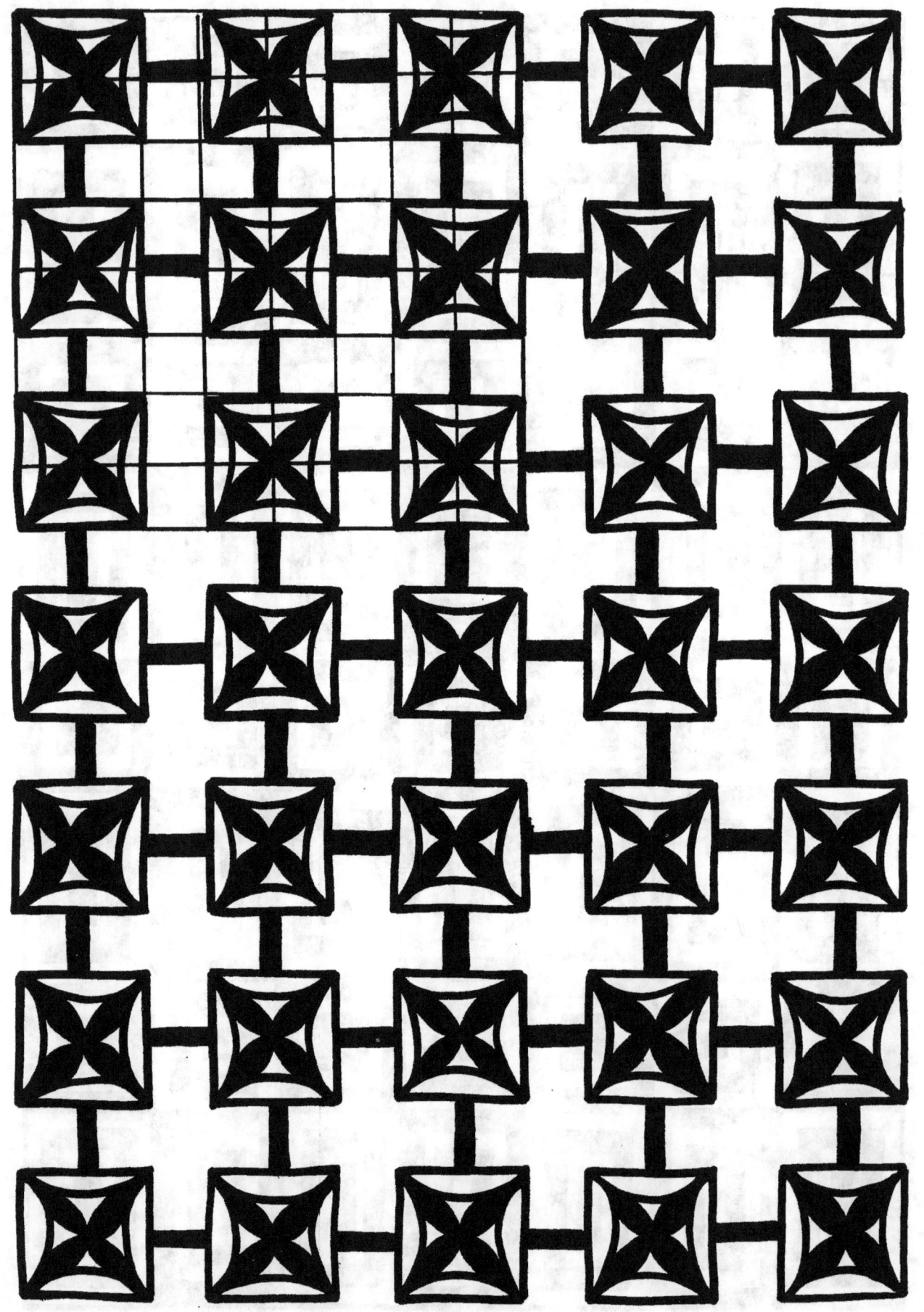

33

34

35

36

37

38

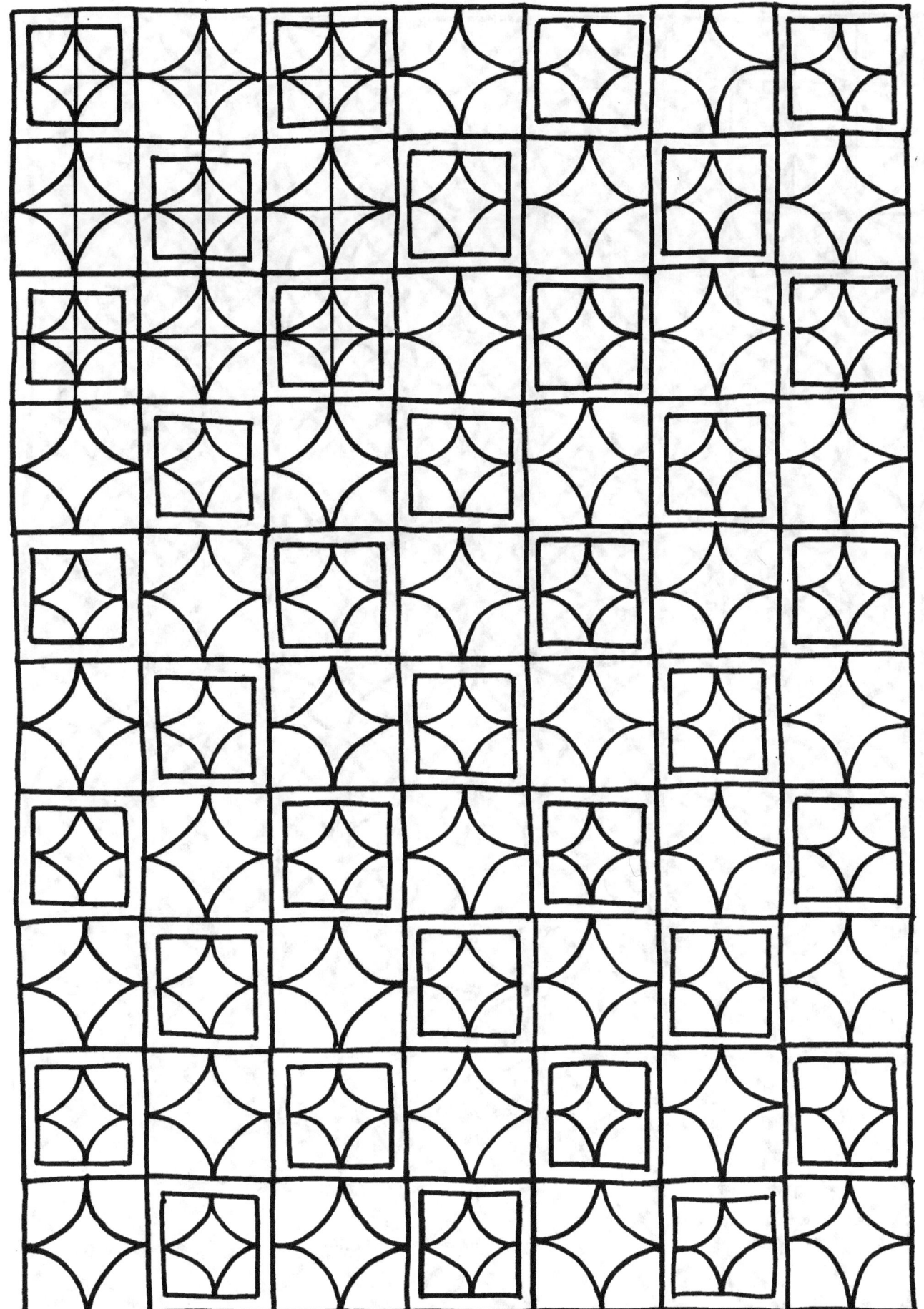

39

40

41

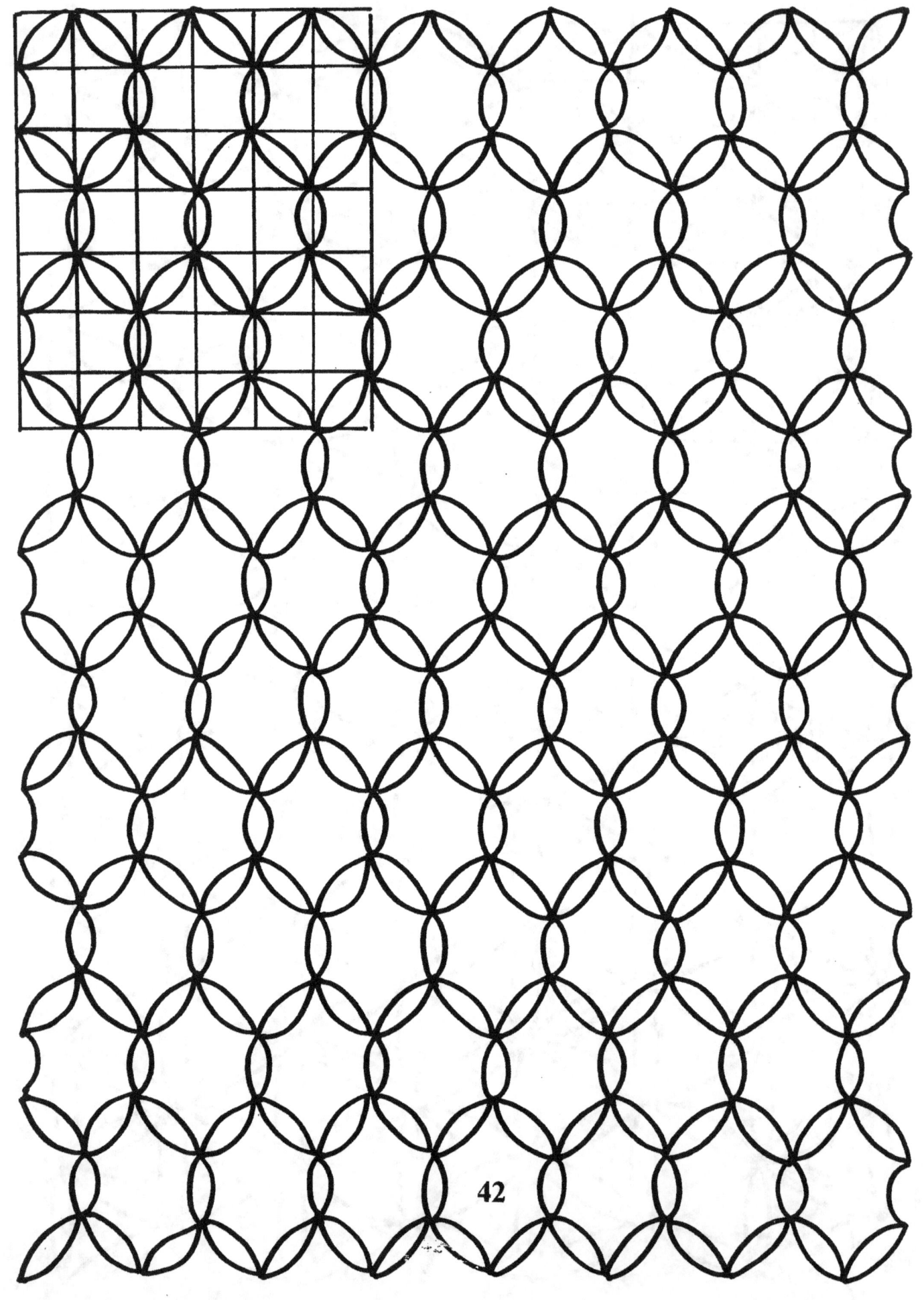

44

47

48

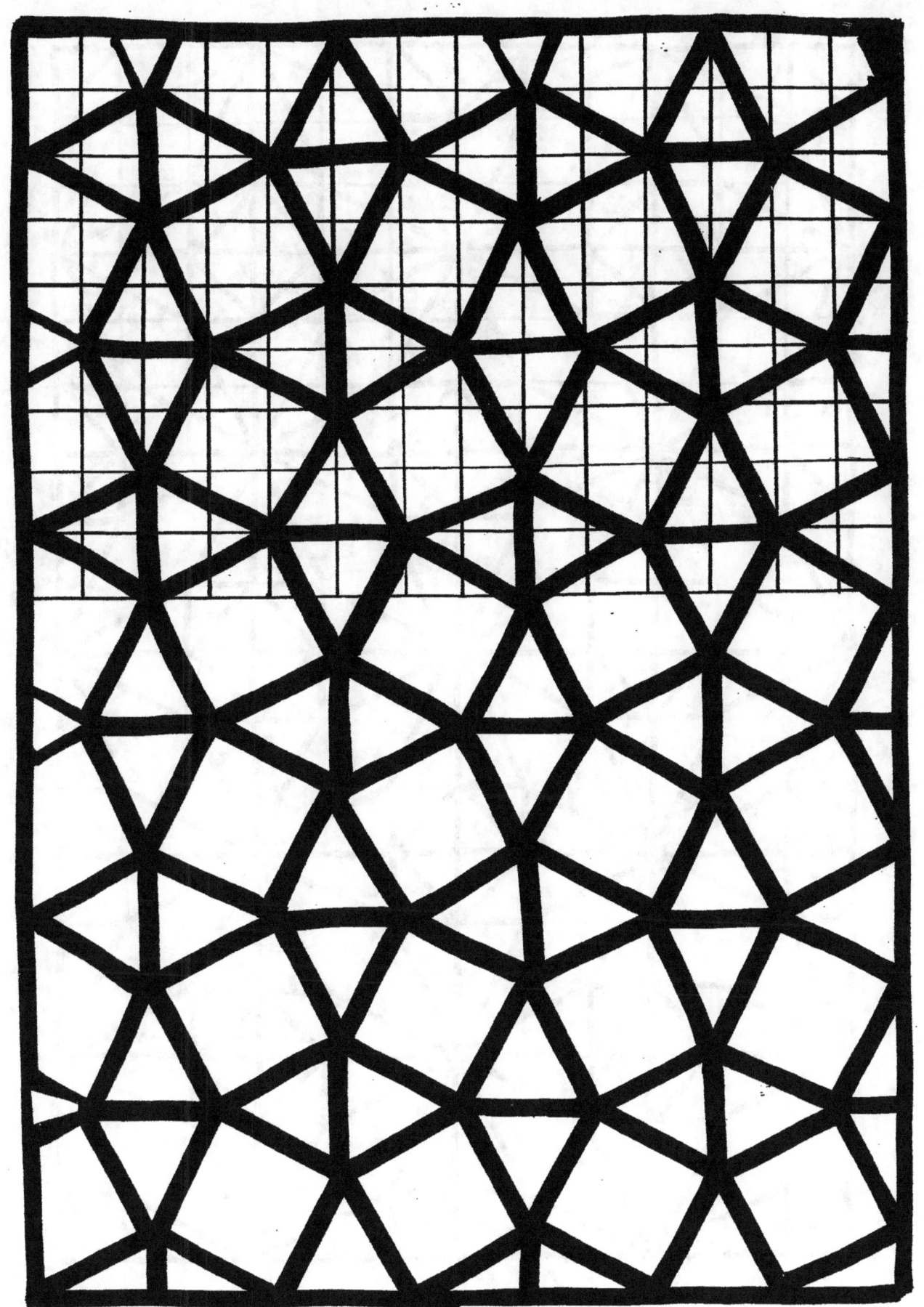

49

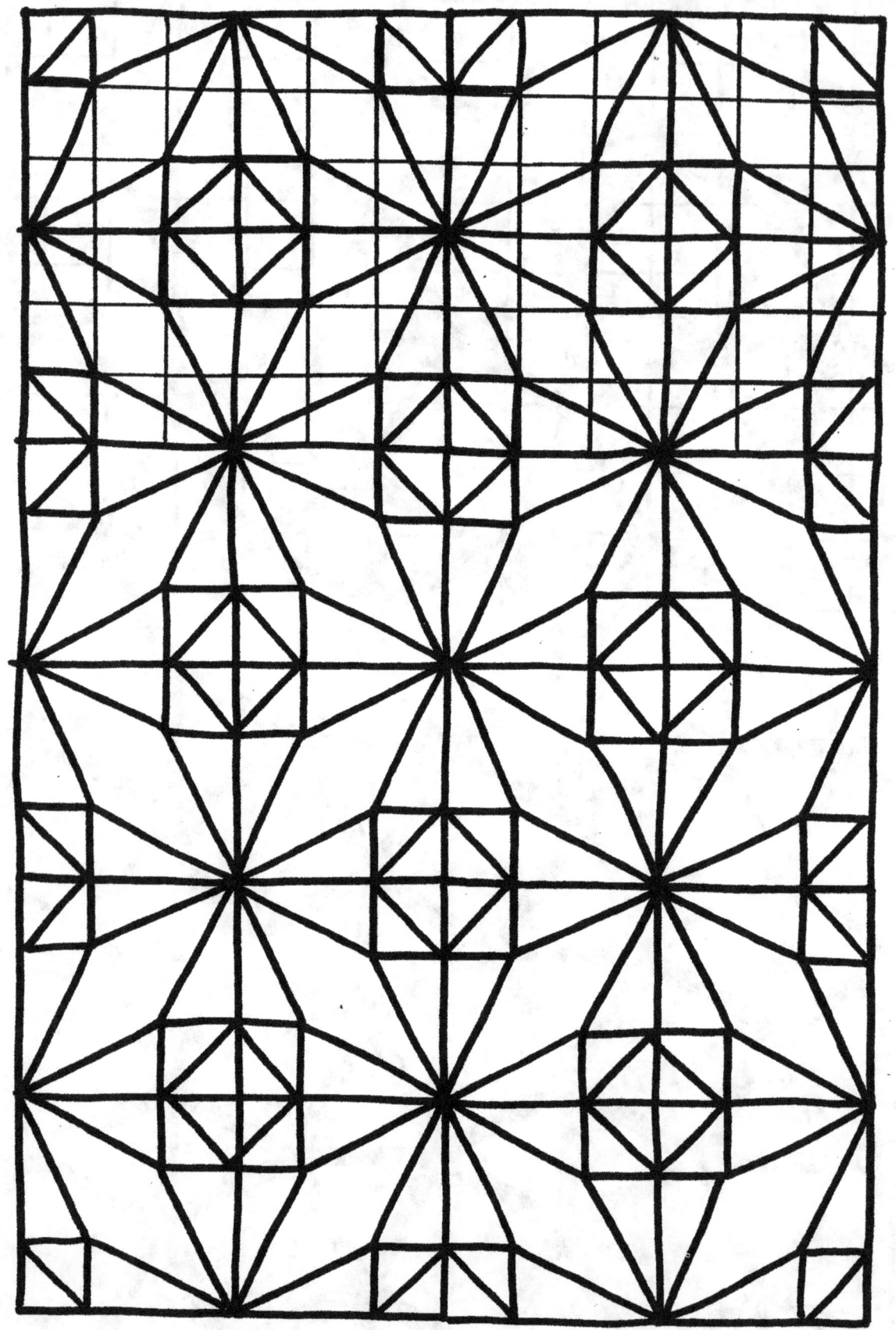

50

52

53

54

57

58

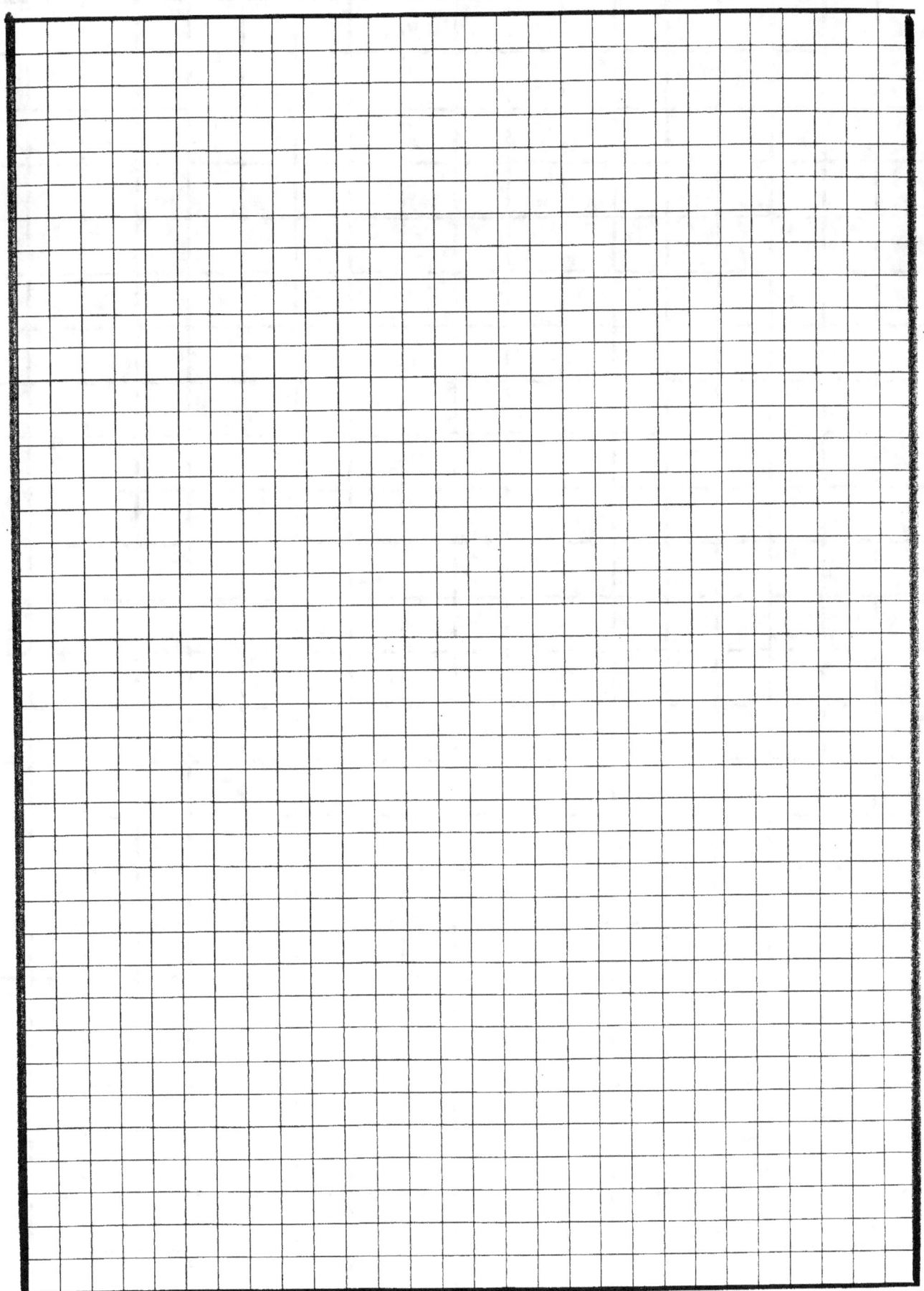

59

60

READING & MATH BOOKS *by JOHN H. LETTAU*

1st Dimension	Grades 3-6
2nd Dimension	Grades 3-6
Primary Dimension	Grades 1-4
Aztec Math Primary Book One	Grades 1-3
Aztec Math Primary Book Two	Grades 1-3
Aztec Math Intermediate Book One	Grades 3-6
Aztec Math Intermediate Book Two	Grades 3-6
Aztec Math Jr. High Book One	Grades 5-8
Aztec Math Jr. High Book Two	Grades 5-8
Aztec Math Decimal Book	Grades 4-8
Aztec Math Fraction Book	Grades 4-8
Sum-Action Number Puzzle Book One	Grades 3-6
Sum-Action Number Puzzle Book Two	Grades 3-6
Sum-Action Number Puzzle Primary Book One	Grades 1-3
Sum-Action Number Puzzle Primary Book Two	Grades 1-3
Multiplication Number Puzzles	Grades 3-6
Geometric Design Puzzle Book One	Grades 3-6
Geometric Design Puzzle Book Two	Grades 3-6
Aztec Reading Primary Book One	Grades 1-3
Aztec Reading Primary Book Two	Grades 1-3
Math in Action	Grades 3-6
A-Maze-ing Number Puzzles	Grades 3-6
Graph Paper Designs	Grades 2-6
Pick-A-Dilly Papers	Grades 3-6
Awards for All Reasons	Grades 1-6
Time Marches On	Grades 1-3
Pennies, Nickels & Dimes	Grades 1-3
Super-Sum Activity Cards	Grades 3-6
Learning Center Game Boards	Grades 1-3
Aztec Design Coloring Book	Grades 1-6

www.ingramcontent.com/pod-product-compliance
Lightning Source LLC
Chambersburg PA
CBHW081609170526
45166CB00009B/2892